Users' Experiences and Challenges with Mobile Financial Services (MFS) in Bangladesh

Users' Experiences and Challenges with Mobile Financial Services (MFS) in Bangladesh

Md. Mehadi Rahman

ELIVA PRESS

ISBN: 978-99949-8-694-1

Cover Design: Eliva Press
Cover Image: Ingimage
Email: info@elivapress.com
Website: www.elivapress.com

This Eliva Press imprint is published by the registered company Eliva Press Global Ltd. part of Eliva Press S.R.L. Publishing Group

The registered company address is: Pope Hennessy Street Level 2, Hennessy Tower Port Louis, Mauritius
Eliva Press S.R.L. Publishing Group address is: Bulevardul Moscova 21, Chisinau, Moldova, Europe

Users' Experiences and Challenges with Mobile Financial Services (MFS) in Bangladesh

Md. Mehadi Rahman

Table of Contents

Chapter One: Users' Experiences of Mobile Financial Services at Rural Areas of Bangladesh

1.1 Abstract

Mobile Financial Services (MFS) is one of the new innovations of the modern financial system. It nowadays plays a crucial role in banking users' day-to-day financial activities. This study investigates users' experiences of MFS at rural areas of Bangladesh. Specifically, this study sought to gain an understanding of to what extent different factors positively or negatively influence users' experience in using mobile financial services. This study was mainly a survey research. The data were obtained using a questionnaire from a total of 128 MFS users (102 male and 26 female) living in rural areas of Bangladesh who were conveniently chosen from five Upazilla in Barishal city. The findings suggested that MFS users in rural areas of Bangladesh have both positive and negative experiences toward the use of mobile financial services. This study further revealed six major factors that have influenced the users' positive and negative experiences in general, i.e. perceived usefulness, perceived ease of use, perceived risk, trust, customer and MFS agent relationship, and perceived cost. It was found that the last factor (perceived cost)have generated the users' negative experiences toward mobile financial services. Therefore, the government and MFS providers are encouraged to work together with some innovative strategies to change the users' negative experiences into positive experiences of using the mobile financial services.

1.2 Introduction

For nearly a decade, rapid development of information and communication technology has significantly affected the banking industry.

Mobile phone is one of the most documented and well-established technologies in developing countries like Bangladesh. Alam (2017) found that only 13% of people have bank accounts in Bangladesh. As mobile phones are cheaply available and mobile operators offer low call rates, most of the people of Bangladesh use mobile phones (Akhtaruzzaman et al., 2017). Recently in Bangladesh, it was observed that traditional branch banking is going to reduce due to the increase in mobile financial services (Bhuiyan & Rahman, 2013). Akhtaruzzaman et al. (2017) suggest that the mobile platform offers a suitable method for managing money without handling cash. Mobile phone operators recognize MFS as a potential service that can be offered to the customers. Alternatively, banks and other financial institutions see MFS as a medium to provide services to the unbanked and rural people of the country. Technological advances and consumer confidence in mobile financial services have made possible the rapid growth of MFS. Mobile financial services can be divided into mobile banking and mobile payment (Bhuiyan & Rahman, 2013).

Mobile financial services are the products and services that a financial institution such as banks provides to its customers through mobile devices. The mobile channel provides a chance for financial institutions to boost customer access to different financial products or services and reduce costs (Akhtaruzzaman et al., 2017). According to Donner and Tellez (2008), mobile banking, m-payments, m-transfers, and m-finance refer jointly to a set of applications that allow people to use their mobile phones to control their bank accounts, store value in an account linked to their mobiles, transfer funds, or even access credit or insurance products.

According to Sultana and Khan (2016), mobile financial services are also known as mobile banking that performs finance associated functions on a mobile device like a smartphone or tablet. With the use of a mobile, users can perform mobile banking via call, text, website, or app. Mobile banking is referred to as actions on a traditional bank account through mobile devices. These actions include obtaining account information, doing banking transactions, and so on through mobile devices. Mobile banking is offered by the most prominent banks. Generally, people use a smartphone app to perform bank transactions securely (Akhtaruzzaman et al., 2017).

Mobile payments are paying for goods and services using a mobile device as the transaction terminal. Mobile payments can use a traditional bank account or an MFS account (Akhtaruzzaman et al., 2017). Branchless banking via a mobile device is also another part of MFS. Customers can take advantage of this service to make basic payments like utilities and other bills, and domestic and international remittances. These transactions become swift, simple, and cost-effective through MFS (Sultana & Khan, 2016). Mobile banking as the channel of branch banking became very helpful to rural customers in saving the travel time and money to visit the distant branches for the financial transaction (Dhillon, 2014).

Bangladesh Bank has introduced mobile financial services in 2011 which promoted only scheduled commercial bank-led MFS. According to Central Bank of Bangladesh (2018), Bangladesh Bank permits the following Mobile Financial Services: (a) disbursement of inward foreign remittances; (b) cash in/out using mobile account through agents/bank branches/ATMs/mobile operator's outlets; (c) person to business (P2B) payments; (d) business to person (B2P) payments;(e) government to person

(G2P) payments; (f) person to government (P2G) payments; (g) person to person (P2P) payments; and (h) other payments like microfinance, overdrawn facility, and insurance premium.

Central Bank of Bangladesh has provided approval to 28 banks for running mobile banking services but at present 15 banks are currently providing MFS in Bangladesh. According to the Bangladesh Bank MFS report, the number of registered MFS clients stood at 92.94 million in August 2020. It means that more than 55% of the total population is under MFS coverage in Bangladesh. The number of active accounts stood at 42.68 million in July 2020. The number of daily transitions and amount of daily transactions stood at 10.01 million and BDT 20,322.4 million in July 2020. Among different services, cash in, cash out, and P2P transaction services dominate the mobile financial activities. The amount of cash in, cash out and P2P transaction stood at BDT 1,70,588 million, BDT 1,91,462.5 million, and BDT 1,72,934.3 million respectively in July 2020.

Customer experience is distinctive and personal to each user. Although any service can be highly consistent and identical, each customer perceived their own experience. User experience creation is two-way. The service provider creates one part of it, and another part will be created and perceived by the customer. Customer experience is defined as being holistic and involving the customer's cognitive, affective, emotional, social, and physical responses to the service provider (Klaus & Maklan, 2013). According to Heinonen, Strandvik, and Voima (2013), customer experience refers to all points of time (before, during, and after the current moment); in addition to the illusion of an ideal service-use experience. Experience is created in the mind of a person who is connected to a service on an emotional, physical, intellectual, or spiritual level. Besides, the experience is subjectively socially constructed, with being contextual and situation-specific (Helkkula & Kelleher, 2010). As experience is socially constructed, the experiences and accounts of other people can affect the customer experience (Dube & Helkkula, 2015). Helkkula and Kelleher (2010) suggest that customers are active producers of both their actual and imaginary experiences.

In the MFS context, where the service is new and developing, imaginary experiences and ideas about the future development of the service are also important and require to be taken into account. The customer experience creates value for the users through how the service is delivered to them and how the interaction between the mobile service and the user functions (Dube & Helkkula, 2015). Interaction in the case of a mobile financial service refers to interaction with the application, which may have a strong influence on customer experience. Alkhaldi (2017) mentioned recent studies in the field of technology acceptance have verified that users' experiences, as well as users' awareness, have not been sufficiently considered. Shaikh and Karjaluoto (2015) reported that despite several factors that have been identified as appropriate factors to be considered for example users' experience and users' awareness; they have not been adequately measured. Different research findings demonstrated that there is a link between users' experiences and the use of m-banking and it is very important to consider users' experience concerning mobile banking use (Amin & Ramayah, 2010; Shaikh & Karjaluoto, 2015).

Globally, mobile banking is one of the latest mobile technology efforts (Shaikh & Karjaluoto, 2015) and so technology-based studies recommend that users' intentions toward m-banking use should be investigated. In several developing countries, customers having more mobile phone experience, ultimately have accepted and using MFS more intensive (Chitungo & Munongo, 2013). Customer satisfaction has a strong relationship with the use of mobile banking. The core factors that determine user experience in mobile banking are service quality, functional quality, perceived value, agent-customer engagement, perceived usability, and perceived risk. There is a significant relationship between user experience, satisfaction, and loyalty, which are related to the financial performance of the financial institution (Mbama, 2018). Customers are probably to use mobile financial services if they perceive these services as reliable with their experience (Kabiry & Forghani, 2013). This provides indirect proof that users' experiences have a significant impact on their awareness and motivation in adopting mobile financial services (Komulainen & Saraniemi, 2019). Medberg and Heinonen (2014) mentioned that customers' needs are growing as more and more transactions take place on mobile phones, demanding real-time, personalized, and seamless payment experiences.

As a result of the above explanation, understanding users' experience regarding mobile banking services has become increasingly important. Moreover, the majorities of mobile banking studies are quantitative and focus on the customers' behavioral intention to adopt mobile banking services (Shaikh & Karjaluoto, 2015). However, recent literature criticizes this and suggests that research should focus more on actual usage behavior, instead of only on use intentions (Laukkanen, 2016). Besides, small number of research have been conducted on mobile financial services especially focusing on the rural people of Bangladesh. As mobile financial services are new for rural people, and so it is very necessary to examine rural users' experience of mobile financial services in Bangladesh.

Drawing on the above gaps, this study shed some light on the users' experiences of using mobile financial services in rural areas of Bangladesh. Using a survey research, this study seeks to answer the question 'how banking users at rural areas of Bangladesh experience the use of mobile financial services?' This study's results potentially contribute to providing insights and references for Bangladesh government and Bangladesh banks to develop and enhance the efficacy of mobile financial service practices, particularly in rural areas of Bangladesh.

1.3 Method

1.3.1 Rationale of the Method

A research design is defined as a plan of research specifying what is to be done and how to do it (Creswell, 2012). This research followed a quantitative approach using survey design. The rationale of survey design is to measuring current beliefs, attitudes, opinions or practices in a short amount of time. In the context of this study, survey research was conducted to examine beliefs and perspectives of mobile financial service users at rural areas of Bangladesh.

1.3.2 Sample Selection and Data Sources

This study was conducted in a small town and village area of the Barishal district of Bangladesh. The population of this study consisted of mobile financial service users who live in that rural area. A sample of 128 MFS users was conveniently selected as participants for conducting the study considering the accessibility of the researcher. The participants were identified from five Upazila (sub-district) which were located in the Barishal district, which is also the researcher's place of living. This allowed the researcher to manage time and collect data within a short period of time. This technique of data collection also kept the cost affordable for the researcher.

1.3.3 Instrumentation and Data Collection

This study used a questionnaire with a Likert type scale to collect users' experiences on mobile financial service usage. The questionnaire consisted of three sections. The first section focused on the respondents' demographic information. The demographic variables included gender, age, occupation, level of education, and monthly income. The second section concerned on the respondents' basic usage of MFS. This section included whether respondents' have a bank account, the name of the MFS they are using, and the time duration of their use. The respondents were also requested to indicate which services they are using regularly, the number of MFS use on monthly basis, the time duration of a transaction, and travel time to the nearest MFS agent. Furthermore, the third section asked the respondents to indicate their agreement or disagreement with the statements based on their MFS using experience.

The questionnaire was 5 points Likert-scale and the anchors vary from "Neutral/don't know" to "strongly agree". The values of anchors were as follows: "Neutral/don't know"= 0, "strongly disagree"= 1, "disagree"= 2, "agree"= 3 and "strongly agree"=4. Here, the researcher made 'zero' as neutral or don't know so that it can be excluded from the calculation and won't interfere with the calculation of user experiences. There were 18 statements in the 5 points Likert-scale of the questionnaire.

The purpose of the questionnaire was to understand the respondents' experience regarding MFS use. The questionnaire was structured and examined users' experiences focusing on users' perceived usefulness, perceived ease of use, perceived risk, perceived cost, agent-customer engagement, and trust towards the MFS.As the respondents were from rural areas and they have low access to the internet, so the questionnaire was distributed manually through offline papers.

1.3.4 Validity and Reliability of the Instrument

The instrument was developed by the researcher based on relevant literature focusing on specific research questions of the study. Initially, related literature was reviewed, and based on that a draft of instrument was developed. Then, initial editing of the instrument was made as per the consultation with the expert on the related field. Finally, the draft instrument was piloted and based on feedback received from the tryout phase and the expert's comment the researcher finalized the instrument.

Cronbach's Alpha was designed as a measurement of internal consistency of items in the Likert type scale of the questionnaire. It varies between zero and one. The closer alpha is to one, the greater the internal consistency of the items in the Likert-type scale of the questionnaire. This study's Likert-type scale consisted of 18 items and Cronbach's alpha test was performed to check the reliability of items. The score of Cronbach's alpha test was 0.783 indicating internal consistency of the items. As the Cronbach's alpha score is >.75, so it can be concluded that the internal consistency of Likert-type scale items is good (Tavakol & Dennick, 2011) and the instrument is reliable.

1.3.5 Data Analysis

The data from the survey were analyzed using a quantitative or statistical approach for answering the research question. The analysis included descriptive techniques as percentages, mean and standard deviation. Initially, the frequency of responses of every item of the questionnaire was categorized. The data was analyzed using SPSS 20.0 for Windows. For 5 point Likert type scale, neutral or don't know was excluded from the calculation. Then, the categories "strongly agreed" and "agreed" were collapsed into one category of "Agreed". Similarly, the categories "strongly disagreed" and "disagreed" were collapsed into one category of "Disagreed". Then the percentage for each category against each response was counted. The data were analyzed based on the weighted mean of statements (Mohiuddin, 2015).After analyzing the data results of survey, findings were presented through tables. The data were analyzed according to the research question and is presented section by section.

1.4 Results

1.4.1 Respondents' Information about Using MFS

1.4.1.1 Users' Information about Bank Account

Figure 1 shows that most of the respondents (78.13%) have bank accounts in several banks of Bangladesh. There was only one-fifth of the respondents (21.88%) have no bank accounts. This 21.88% of respondents only have mobile financial service accounts. They are not very familiar with formal banking services and so they avoid opening any bank accounts. In contrast, as they could easily use MFS with their mobile phone, so they became accustomed to mobile banking.

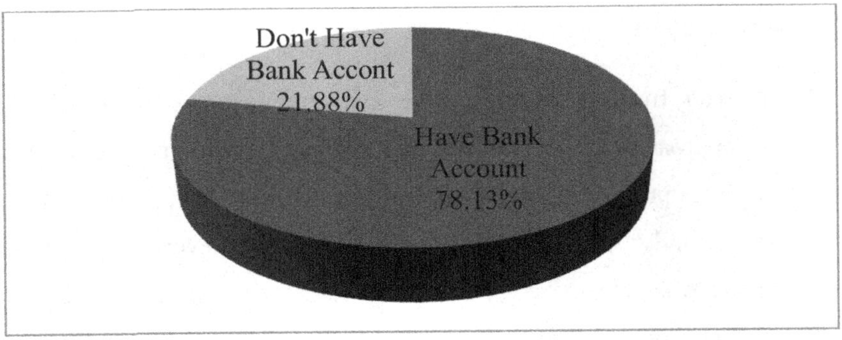

Figure 1. Respondents' Information about Bank accounts

1.4.1.2 Mobile Financial Services Used by Respondents

Figure 2 describes that 84.38% of the respondents are using the most popular mobile financial service in Bangladesh named "Bkash". Half of the respondents are using "Rocket" which has initially introduced MFS in Bangladesh. Only 14.06% of the respondents are using "Nagad" which is currently growing very its popularity in Bangladesh.

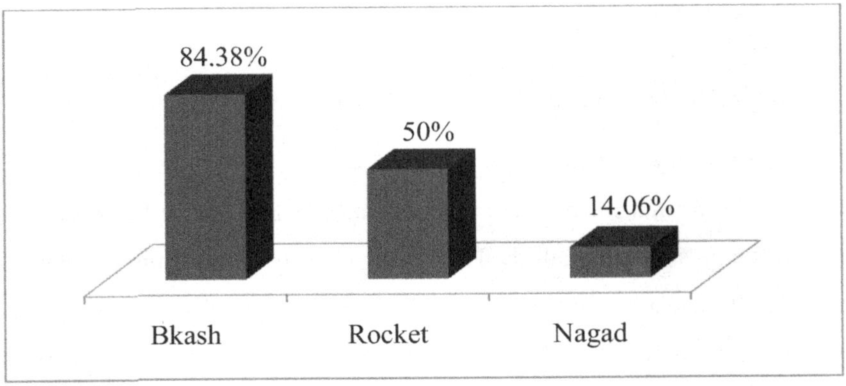

Figure 2. Percentages of users who use MFS

1.4.1.3 Length of MFS Use

Table 1 demonstrates that most of the respondents (31.25%) are using MFS for more than 5 years. The highest percentage of respondents (28.13%) are using MFS for 3 to 5 years. Some respondents (18.75%) are using MFS for 1 to 2 years, while only few respondents (12.5%) are using it for 2 to 3 years. The rest 3.13% of respondents are new MFS users (using for less than six months) and 6.25% are using for 6 months to less than one year.

Table 1. Length of MFS Use

Time Range	**Number of Users**	**Percentages**
>6 months	4	3.13%
6 months - 1 year	8	6.25%
1 year - less than 2 years	24	18.75%
2 years - less than 3 years	16	12.50%
3 years - less than 5 years	36	28.13%
More than 5 years	40	31.25%

1.4.1.4 Different Services of MFS

Table 2 portrays that more than two-thirds of respondents are regularly using cash in (68.75%), cash-out money from agents (71.88%), sending money to others (70.31%), and recharging mobile balance (79.69%) through their MFS. Some respondents are also checking balance statements (40.63%) and cash-out money from ATM (32.81%) regularly. Few respondents are using MFS for bill payment (18.75%), transfer money within bank and MFS (9.38%), and getting international remittance (1.56%).

Table 2. Different Services of MFS

Time Range	Number of Users	Percentages
Cash in	88	68.75%
Cash-out from agent	92	71.88%
Cash-out from ATM	42	32.81%
Send Money	90	70.31%
Mobile Recharge	102	79.69%
Bill Payment	24	18.75%
Balance check/Statement	52	40.63%
Transfer Money (Bank to MFS/ MFS to Bank)	12	9.38%
International Remittance	2	1.56%

1.4.1.5 Monthly Frequency Use of MFS

Figure 3 shows that the majority of the respondents (37.50%) are using MFS 4 to 8 times a month. Some respondents (26.56%) are using MFS less than 3 times in a month. Few respondents (14.06%) use MFS 9 to 15 times a month. There are only 10.94% of the respondents use MFS either 16 to 20 times or more than 20 times.

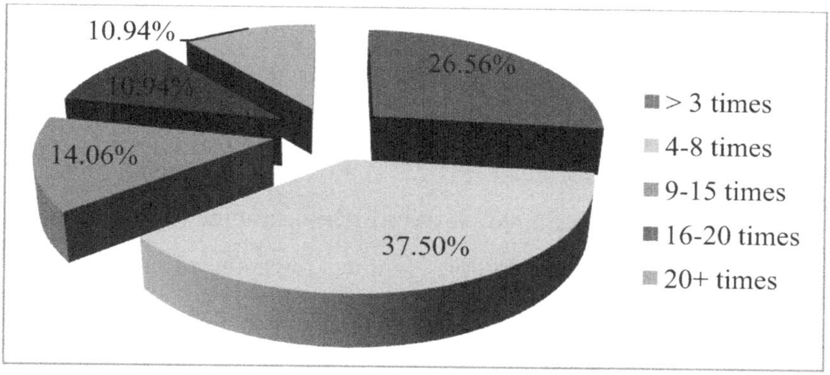

Figure 3. Frequency of MFS Use in a Month

1.4.1.6 Transaction Time for Using MFS

Figure 4 reveals that most of the respondents (62.5%) needed one to three minutes for making a transaction. Some respondents (20.31%) experienced that in less than a minute they could do a transaction with MFS. Few respondents (17.19%) also found that the transaction time was around four to six minutes.

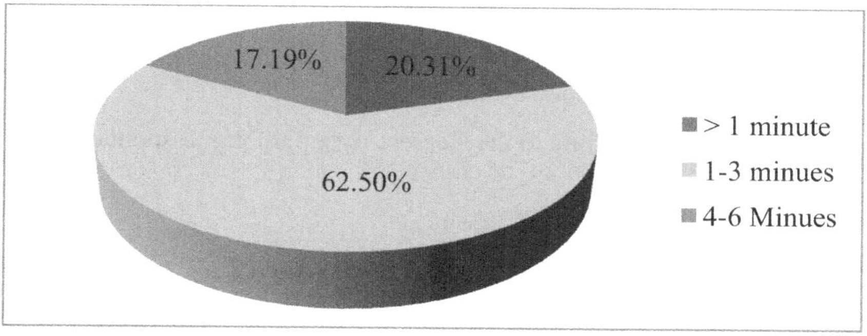

Figure 4. The Transaction Time for MFS Users

1.4.1.7 Respondents' Travel Time to the Nearest MFS Agent

Figure 5 implies that more than one-third of the respondents (39.06%) informed that travel time to the nearest agent is less than five minutes. Almost one-third of respondents (31.25%) travel time is five to ten minutes. 12.5 % of respondents required eleven to twenty minutes and 9.38% of the respondents required twenty to thirty minutes for going to the nearest MFS agents. Only 7.81% of the respondents needed more than thirty minutes for travelling to the nearest MFS agent.

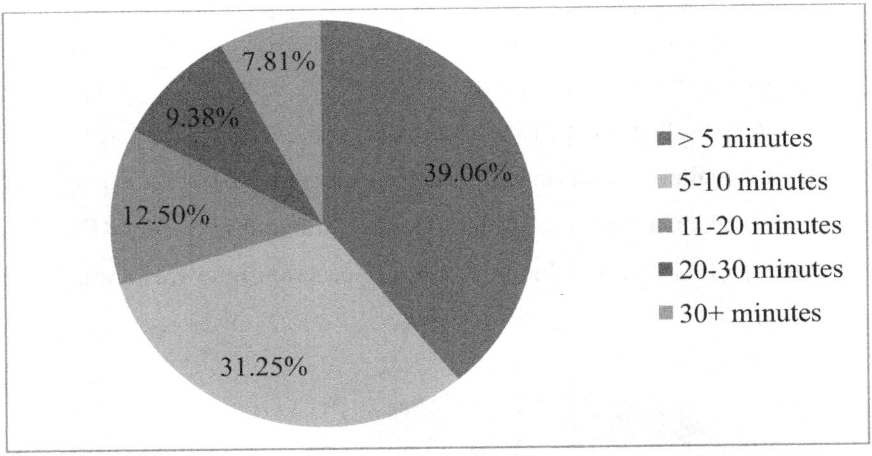

Figure 5. Travel Time to the Nearest Agent among Respondents

1.4.2 Results of Data Analysis

The analysis of the Likert scale results was conducted by determining frequency, mean, and percentage. In this analysis, statements that are correlated to one another are put in the same table with their underlying theme as the section heading. A criterion was developed to interpret based on the weighted mean of each statement of the experience.

Table 3. Scaling of Weighted Mean

Class Interval (Weighted Mean)	Categories
1.00-1.75	Strongly Disagree
1.76-2.50	Disagree
2.51-3.25	Agree
3.26-4.00	Strongly Agree

1.4.2.1 Users' Experiences on the Usefulness of MFS

Here, table 4 consists of statements congruent with the perception of MFS usefulness. As shown in Table 4, most of the respondents (96.9%) agreed with the idea that MFS helped them to make payments more quickly (M=3.09; SD= .46). Similarly, a large proportion of the respondents (93.8%) agreed that MFS has enhanced their payment effectiveness (M=3.06; SD=.50). More than 90% of the participants perceived that MFS is very useful (M= 3.09; SD= .64). In terms of mean, the respondents agreed to all the statements. Therefore, it can be concluded that in general the respondents agreed (WM=3.08) that MFS is very useful for them.

Table 4. Users' Experiences on Perceived Usefulness of MFS

Statements	Percentage		Total	Mean	SD	WM
	Disagree	Agree		(M)		
1. MFS helped me to make payments more quickly.	3.1%	96.9%	100%	3.09	0.46	
2. MFS has enhanced	6.2%	93.8%	100%	3.06	0.50	**3.08**

my payment effectiveness.						
3. MFS is very useful.	9.3%	90.7%	100%	3.09	0.64	

Note: SD= Standard deviation, WM= Weighted mean

1.4.2.2 Users' Experiences on the Perceived Ease of Use of MFS

Table 5 demonstrates that the statements congruent with the perception about ease of use of the MFS. As shown in Table 5, more than 90% of the respondents agreed that MFS is very easy to use (M=3.17; SD=.61) and the payment process is very clear and understandable for them (M=3.03; SD=.53). Quite a large portion of respondents (71.9%) agreed that the MFS payment process is very simple which needs no assistance (M=2.84; SD=.72). But some respondents (28.1%) disagreed with this statement. In terms of mean, the respondents agreed to all the statements related to ease of use about MFS. Therefore, it is concluded that the respondents experienced (WM= 3.01) the easy use of MFS.

Table 5. User perceived ease of use about MFS

Statements	**Percentage**		**Total**	**Mean**	**SD**	**WM**
	Disagree	**Agree**		**(M)**		
1. MFS is very easy to use.	4.7%	95.3%	100%	3.17	0.61	
2. During MFS payment, the process is very clear and understandable.	9.3%	90.7%	100%	3.03	0.53	**3.01**

3. MFS payment process is very simple that assistance is not needed.	28.1%	71.9%	100%	2.84	0.72	

Note: SD= Standard deviation, WM= Weighted mean

1.4.2.3 Users' Experiences on the Cost of MFS

Table 6 consists of statements congruent with respondents' experiences on the cost of using MFS. As shown in Table 6, more than two-thirds of the respondents (73.4%) perceived that the transactional charge of MFS is not affordable (M= 2.16; SD= .77). In the contrary, some respondents (26.6%) agreed to the idea that the transactional charge of MFS is affordable. Most of the respondents (70.3%) disagreed that the cost of MFS is lower than manual banking transaction (M= 2.11; SD= .86). But some respondents (29.7%) agreed with this statement. Hence, it is concluded that in terms of mean, the respondents disagreed to all the statements related to the cost of MFS. In other words, they experienced high cost of using MFS (WM= 2.14).

Table 6. User perceived cost about MFS

Statements	**Percentage**		**Total**	**Mean**	**SD**	**WM**
	Disagree	**Agree**		**(M)**		
1. Transactional charge of MFS is very affordable.	73.4%	26.6%	100%	2.16	0.77	**2.14**
2. Cost of MFS is	70.3%	29.7%	100%	2.11	0.86	

lower than traditional
banking.

Note: SD= Standard deviation, WM= Weighted mean

1.4.2.4 Users' Experiences on the Trust on MFS

Table 7 consists of statements congruent with the respondents' trust to the MFS. As shown in Table 7, nearly two-thirds of the respondents (65.6%) perceived that the MFS system is very trustworthy (M= 2.61; SD= .73). However, more than one-third of the respondents (34.4%) disagreed with this statement. Although a large portion of respondents (61%) agreed that MFS always provides safe and secure services (M= 2.58; SD= .79), a large number of respondents (39%) disagreed with this statement. Most of the respondents (79.7%) experienced that MFS provides error-free records (M=2.78; SD=.58). Thus, in terms of mean, the respondents agreed with all the statements in case of trusting MFS. So, it is concluded that MFS can be trusted (WM= 2.66).

Table 7. Users trust towards MFS

Statements	**Percentage**		**Total**	**Mean (M)**	**SD**	**WM**
	Disagree	**Agree**				
1. MFS system is very trustworthy.	34.4%	65.6%	100%	2.61	0.73	
2. MFS always provides safe and secure services.	39%	61%	100%	2.58	0.79	**2.66**
3. MFS provides error free records.	20.3%	79.7%	100%	2.78	0.58	

Note. SD= Standard deviation, WM= Weighted mean

1.4.2.5 Users' Experiences on the Perceive Risk of MFS

Table 8 consists of statements congruent with the respondents' perceived risk with MFS. As described in Table 8, more than half of the respondents (56.3%) perceived that the MFS system provides risk-free transaction service (M= 2.63; SD= .70). However, nearly half of the respondents (43.7%) disagreed with this idea. Moreover, more than 60% of the respondents agreed that MFS has a low risk of losing their money or financial details (M= 2.52; SD= .76) and also low uncertainty with MFS payment (M= 2.62; SD= .66). Around 40% of respondents showed disagreement with the statements. In terms of mean, the participants agreed to all the statements regarding perceived risk with MFS. Therefore in general the respondents perceived that MFS has low risk (WM=2.59).

Table 8. Users perceived risk with MFS

Statements	Percentage		Total	Mean	SD	WM
	Disagree	Agree		(M)		
1. MFS system is risk-free transactions than other financial services	43.7%	56.3%	100%	2.63	0.70	
2. There is a low risk of losing my money/financial details from MFS account.	39.1%	60.9%	100%	2.52	0.76	**2.59**
3. There is low uncertainty associated with using MFS	37.5%	62.5%	100%	2.62	0.66	

payment.

Note: SD= Standard deviation, WM= Weighted mean

1.4.2.6 Users' Experiences on MFS Agent-Customer Engagement

Table 9 presents the respondents' experiences with MFS agent-customer engagement. As shown in Table 9, more than half of the respondents agreed that MFS agents deal their complaints with care (M= 2.58; SD= .75) and also provide solutions within a limited time (M= 2.45; SD= .62). On the contrary, around 50% of respondents disagreed with these statements. A number of 60% of the respondents experienced that MFS agents always serve them with a smile (M= 2.56; SD= .61). However, 40.6% of the respondents disagreed with this idea. Most of the respondents (68.8%) experienced that MFS agents are very available for them (M= 2.67; SD= .67). Therefore, it can be concluded that in general the respondents agreed that MFS agent-customer engagement is positive (WM= 2.57).

Table 9. Users experience regarding MFS agents customer engagement

Statements	**Percentage**		**Total**	**Mean (M)**	**SD**	**WM**
	Disagree	**Agree**				
1. MFS agents deal customer complaints with due care.	45.3%	54.7%	100%	2.58	0.75	
2. MFS provides complete solution to individual needs in time.	48.4%	51.6%	100%	2.45	0.62	**2.57**
3. MFS agents always	40.6%	59.4%	100%	2.56	0.61	

serve people with smile					
4. MFS agents are very available for people.	31.3%	68.8%	100%	2.67	0.67

Note: SD= Standard deviation, WM= Weighted mean

1.5 Discussion

1.5.1 Rural Users' Usage of Mobile Financial Services

This study revealed that most of the rural MFS users (78.13%) have bank accounts. Bkash is the most dominant MFS brand used by nearly 85% of the rural users. Bkash is the most popular mobile financial service in Bangladesh (Islam et al., 2017). Mobile banking started the journey to bring the poor under the umbrella of the banking sector especially the rural poor as there are not many bank facilities, and their savings are low so they feel uncomfortable to go to the bank (Alam, Patwary & Rahim, 2013). MFS provides to benefit millions of rural people in Bangladesh by giving them financial independence on top of making life comfortable (Hossain & Russel, 2017). Rural people were introduced to the formal banking system through mobile banking. The use of MFS has paved its way to open bank accounts. This study's result indicates that rural users are getting included in the formal banking system.

It is found that nearly 60% of rural users are using MFS for more than at least 3 years. MFS has become very popular for the last couple of years and it has also covered the rural areas. It is because MFS not only offers a safe and convenient mechanism for saving but also makes an encouraging environment for women to save more money (Hossain & Russel, 2017). Additionally, no cost is needed for having an MFS account which has attracted the un-banked population of Bangladesh (Rahman et al.,

2020). For the last couple of years, the number of MFS users is rising gradually. The concept of digitalizing the whole country has also covered many spheres of rural people's life. The increase in technology usage in the urban areas is influencing user behavior, payment techniques, and purchasing habits too (Akhter & Khalily, 2020). The more rural people are using MFS; they are getting more comfortable with MFS.

Cash in, cash out, sending money and mobile recharge are some most common and frequent services that rural users are using with their MFS. Sarpong and Agbeko (2020) found that airtime top-up and sending and receiving money were the two most common transactions that were frequently done by users. Hasan (2020) also found that the most common use of mobile banking is balance transfer between the users and users transfer their balance for different purposes. MFS started its journey with features like sending and receiving money among rural users. Still, now rural users are frequently using these services. Chavali and Kumar (2018) expressed that mobile banking offers a wide range of services for users to choose from. However, rural users have limited knowledge about different new features of MFS like merchant payments or online shopping. Akhter and Khalily (2020) mentioned that MFS is a simple, cost-effective, safe, and affordable way to send and receive money. Rural women use MFS mainly for receiving money from their husbands working in urban areas. Most of the time rural people transact small amounts of money within their limits (Hossain & Russel, 2017).

This study verified that majority of the rural users (37.5%) either use MFS four to eight times or more than nine times (35.9%) monthly. The frequency of using MFS is increasing rapidly among rural users. Mbama

(2018) argued that users who frequently use digital banking are enjoying it to use it for their good experience with it. On the contrary, Hossain and Russel (2017) mentioned that at present savings tendency through MFS is not in enormous volume due to the lack of facility, lack of product diversification, and business operation duration of mobile financial service. Rural users prefer to get hard cash rather than saving digital money in their MFS accounts. Hasan (2020) claimed that MFS payment is a popular method that is also a part of modern e-transactions for favorable offers by mobile banking outlets (especially shops or supermarkets). Rural people will be encouraged to use MFS if they get these types of payment facilities in their respective areas. Even if payment offers are unavailable in rural areas, utility bill payment (electricity, gas, telephone, etc.) is growing up usage among users in Bangladesh. MFS has created a positive experience among users because users believe that all services of MFS are very good for them (Alam et al., 2013).

This study shows that the maximum number of rural users' transaction time was one to three minutes. Nearly 70% of rural users could reach their nearest MFS agent within ten minutes. Similarly, Dona et al (2014) also found that there were available agents and branches of DBBL mobile banking, Bkash, mCash, and so on. Additionally, Sarpong and Agbeko (2020) found that most of the users claim it takes not more than 5 minutes during a transaction and not more than 15 minutes to travel to an agent. Rural users can easily reach their MFS agents and get the necessary mobile banking service from them. As transaction time is shorter, so users feel more comfortable making a transaction through MFS. Besides, Alam et al. (2013) found that users believe MFS transaction is a speedy process because it can be done anytime anywhere quickly in less time. People believe that MFS takes less time than traditional banking (Ahmed et al., 2012). People don't have to wait in the queue or fill up different papers to transfer money. MFS has reduced the burden of paperwork and time consumption of traditional banking for rural people. So, everyone believes that MFS is more beneficial due to saving time for them (Sarpong & Agbeko, 2020; Khan et al., 2017). Nevertheless, sometimes users face problems due to late payments or for any other purpose, which causes loss of time (Rahman et al., 2020).

1.5.2 Users' Experiences of Perceived Usefulness

Perceived usefulness refers to the degree to which an individual feels that his/her performance will improve as a result of using a particular system (Davis, 1989).This study finding suggests that rural users have a positive experience regarding the usefulness of MFS. More than 90% of users perceive that MFS is very useful for them because it has helped them to make payments more quickly and also enhanced their effectiveness of payment. Likewise, Komulainen and Saraniemi (2019) found that speed is one of the vital properties of mobile banking for the users, and so they use the service and found it useful. Liza (2014) also found that current users of mobile banking services perceived mobile banking as useful. People will accept mobile banking services when the value and benefit of mobile banking are evident. Continuous development of MFS has made it faster for mass people. Unbanked and rural people were also able to make payments within a few minutes.

Prior research found that perceived usefulness is positively related to system usage (Talukder, 2014). Furthermore, Zarmpou, Saprikis, Markos, and Vlachopoulou (2012) also established the strong influence of perceived usefulness on the adoption of mobile services. The strong, direct effect of perceived usefulness on an intention to use m-commerce has also been confirmed by Chong (2013). Finally, when an individual perceives that innovation offers a relative advantage over the firm's current practice, it is more likely to be adopted and implemented (Talukder et al., 2014). According to Chavali and Kumar (2018), mobile banking helps people in two ways. People can make proper financial planning as they can continuously monitor their financial transactions. Mobile banking also saves time since people need to not to visit the bank branch for a regular transaction like traditional banking. At present many workers get their salary in their MFS account and they regularly make a transaction through MFS. It gave rural people the power to analyze their financial activities frequently. Moreover, MFS reduces the expenses of stepping in and out in addition to allow people living in rural areas to get money in real-time (Hossain & Russel, 2017). Rural users' perceived usefulness has increased their MFS usage and created a positive experience among them.

1.5.3 Users' Experiences of Perceived Ease of Use

Perceived ease of use refers to "the degree to which a person believes that using a particular system would be free of effort" (Davis, 1989, p. 320).This study established that most of the rural users experienced that MFS is very easy to use because the process is very clear and understandable. Over the years providers are trying to make their payment process as simple as they can so that people can do it very easily. Maximum rural users can make a payment without any assistance but some users experience the payment process as complex and need assistance to do it. The previous studies found that current users of mobile banking services perceived mobile banking to be easy to use, all tasks can be accomplished and it does not need a lot of effort (Chavali & Kumar, 2018; Liza, 2014).Chavali and Kumar (2018) added that, if mobile banking technology is easy to use, it will enhance the usefulness of mobile banking and as a result, people will adopt the mobile banking. When people can understand the whole process of MFS simply and do it instantly by themselves, then it creates a positive experience for them.

Moreover, Hossain and Russel (2017) claimed that rural people are getting money through MFS very easily without any harass and can use that money quickly for their convenient time, which helps much for the development of a living standard of rural people of Bangladesh. It is important for users to feel that they know how to use and control the mobile banking service smoothly. The sense of control is affected by the ease, transparency, and logic of the service (Komulainen & Saraniemi, 2019). Rural people never got control over their financial stability until MFS has come into the picture. They can manage their money effortlessly as they can control their MFS account without other help. This study result reflects that easy usability has increased the rate of usage of MFS for all users. Previous research findings also indicated that the perceived ease of use variable has a significant positive relationship with perceived usefulness (Chavali & Kumar, 2018; Liza, 2014). Perceived ease of use has been used as a powerful factor in new technology adoption in mobile services (Zarmpou et al., 2012) and mobile banking (Wei et al. 2009).

1.5.4 Users' Experiences of Perceived Cost

According to Luarn and Lin (2005), the cost is defined as the extent to which an individual believes that using mobile banking would cost money. This study demonstrated that most of rural users were unhappy due to the high transaction cost of MFS. Poor and unbanked rural people don't have the affordability to pay this price. As most of the users have a bank account, they consider that cost of traditional banking is lower than MFS. So, some rural people prefer traditional banking rather than MFS in terms of cost.

The previous studies also found that most of the users are dissatisfied with the high transaction charges of MFS (Hasan, 2020;Sarpong & Agbeko, 2020). In contrast, Alam et al. (2013) found that most of the users believe mobile banking has a lower cost compared to traditional banking. But some users perceived that cash out via agents takes more cost than traditional banking. Wei et al. (2009) claimed that cost can impede the successful development of mobile banking. Chong (2013) suggested that many mobile users are young customers for instance students, and they may be more price-sensitive compared to other consumers. Most of the rural users feel discouraged to use MFS due to its 1.75% cash out charge. The high charges created a negative experience among many rural users. Wessels and Drennan (2010) found that there is a negative relationship between perceived cost and the intention to use mobile banking. It means the higher are the costs of using mobile financial service, the less it will be used by people.

1.5.5 Users' Experiences on Trustworthy of Mobile Financial Services

According to Khan et al. (2017), trust refers to a user's belief that his/her privacy regarding personal and transaction information will not be harmed, and the technology is reliable enough to protect his/her private information from illegal parties such as hackers.This study explored that, although most of the rural users trust the MFS system, more than one-third of users don't trust MFS. The finding suggests that users' experience is quite similar in the case of providing safe and secure services. Similarly, the previous study found that numerous people still do not trust the MFS system as they are concerned about their faulty transactions or frauds (Ahmed, Imtiaz & Kausar, 2020; Ramdhony & Munien, 2013). On the contrary, Malaquias and Hwang (2016) observed a positive association between trust and mobile banking use considering respondents from Brazil and the USA.

Trust can also enhance users' intention to use and increase their perception related to mobile banking satisfaction (Sharma & Sharma, 2019). Trust can help users to overcome perceptions related to insecurity and risks. Trust also affects users' continuance intention in mobile payment (Zhou, 2013). Koksal (2016) claimed that, when trust is developed among people, it reduces perceived risk and uncertainty with mobile banking, thus leading to a favorable effect on customer intentions. Malaquias and Hwang (2016) considered that the frequent use of mobile banking contributes to an increase in trust in mobile devices. So, previous research suggested that banks and service providers should constantly innovate and offer better safe and trustworthy mobile applications to boost users' confidence towards mobile banking services (Chitungo & Munongo, 2013).

1.5.6 Users' Experiences of Perceived Risk

Perceived risk refers to the five facets of risk including performance risk, security/privacy risk, time risk, social risk, and financial risk (Kabir, 2013). If a person perceived higher risks and insecurity for instance issues of loss and theft of money or financial information due to system hacking, this would discourage the adoption of mobile banking by the rural communities (Chitungo & Munongo, 2013). Although most participants agreed that MSF is safe and has low risk, this study results in nearly 40% of the total participants experienced that the MFS system is risky than other financial services and they fear losing their money or financial details from the MFS account. These have generated a high uncertainty to make payment with MFS.

Every mobile financial service is associated with some risk. Moreover, many people fear financial loss because of the hacking of account information. On the contrary, other researchers found that most people believe that mobile banking is fairly or highly secured (Alam et al., 2013; Bhuiyan & Rahman, 2013). If a person perceived higher risks and insecurity for instance issues of loss and theft of money or financial information due to system hacking, this would discourage the adoption of mobile banking by the rural communities (Chitungo & Munongo, 2013). This research suggests that rural people perceive much risk connected with MFS and it generates negative experiences for them. Similarly, Makanyeza (2017) found that perceived risk had negatively influence users' behavioral intention to adopt mobile banking services.

1.5.7 Users' Experiences of Customer and Agent Relationship

Most of the customers seem to face different issues related to mobile banking and so they seek help from mobile banking employees or agents. Customers also file complaints against different problems they face with mobile financial services. Karatepe and Aga (2016) expressed that bank employees have interactions with customers which are the most significant link in service delivery and complaint handling processes. They need to be friendly, proficient, capable of sustaining interpersonal distance (Garg,

Rahman & Qureshi, 2014), and in building trust and influencing customer behavior (Mbama, 2018).

This research showed that almost half of rural users have negative experiences regarding agents' customer engagement. They experienced that agents don't deal with their complaints with due care and took huge time to provide any solution for them. Most of the rural users experienced that agents serve them with a smile and they are always available. Nevertheless, almost one-third of users' experienced agents are unavailable and they are not very friendly at all. Customer care representatives can't give them proper solutions within time. Akhtaruzzaman et al. (2017) explored that the agents are not competent in dealing with customers' queries and complaints. They do not know how to provide entire services to clients. They are even blamed for colluding with miscreants in illegal transactions through MFS in Bangladesh. Mbama(2018) mentioned employee-customer engagement influences and improves customer experience on digital banking. This research shows that sometimes the negligent behavior of agents and customer care representative has formed a negative experience among rural users. MFS agents' engagement with customers may influence their satisfaction and experience with MFS as well as increase profit for a service provider.

1.5.8 Factors Affecting Rural Users Experience with MFS

Finally, this research has established that rural people's user experience regarding mobile financial services is influenced by several factors. Rural users perceived that MFS is very useful and the using process is quite simple for them. Rural users have generated positive experiences due to perceived usefulness and perceived ease of use. On the other hand, rural users have negative experiences due to perceived risk, perceived cost, and trust. Malaquis and Silva (2020) found that ease of use, perceived usefulness, and trust had a positive significant relationship with the use of mobile banking diffusion in rural areas. This research confirmed that unsupportive behaviour from agents and customer care representative has reduced user intention to use MFS. Khan et al. (2017) assumed that the adoption and usage of mobile banking in Bangladesh is reliant on a variety of variables like – convenience, network availability, complexity, security, and trust.

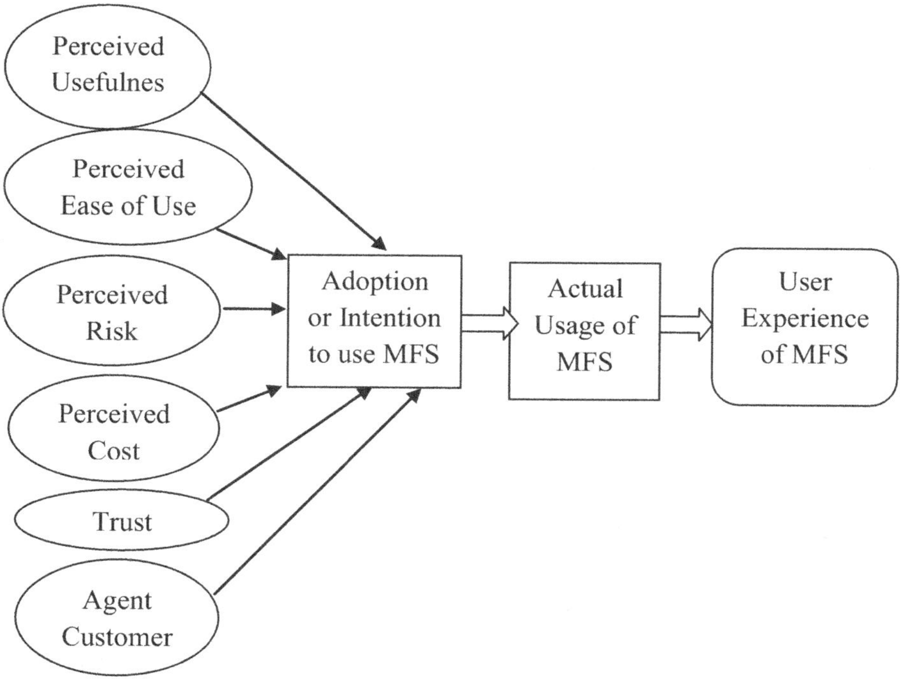

Figure 6: Factors affecting rural users experience with MFS

The result of the study proved that the factors of perceived usefulness, perceived ease of use, perceived risk, perceived cost, agent-customer engagement, and trust are influencing users' adoption or intention to MFS. The more users would use mobile financial services, the more they get used to it. Rural users' intention to use has impacted their actual use of mobile financial services. The main study results indicated that rural users have both positive and negative experiences regarding MFS. These negative experiences of users need to be removed with some preventive measures by MFS providers, the central bank of Bangladesh, and the Government.

1.6 Conclusion

This study aims to shed some light on users' experiences of mobile financial services at rural areas of Bangladesh. This study further examine the users' understanding on to what extent several factors (perceived usefulness, perceived ease of use, perceived cost, trust, perceived risk, and customer-agent relationship) contribute to their perception toward mobile financial services. Drawing on a survey research, this study's findings portray that these factors become considerations of the users both in terms of positive and negative experiences. Although the users provide a fair perception to the majority of above-mentioned factors, it was found that high cost of using MFS partially contributes to their negative view on the mobile financial services offered by several Bangladesh banks. This study indicates that the more users are willing to adopt mobile financial service, the more they get used to it will be. In addition, it is also noticed that their intention to use the mobile financial service has affected their actual use of it in day-to-day financial transaction.

These findings suggest that there is a plethora of scopes where all stakeholders have to work hand-in-hand cooperatively to make MFS more lucrative, particularly for users in rural areas. Only then MFS will contribute to alleviate poverty especially at rural areas and ensure financial inclusion and economic growth of a developing country like Bangladesh. Apart from these compelling results, this study have several limitations. The data were collected within a short period of time due to the bad situation of covid-19. Additionally, The target population and sample size was very small which cannot be generalized for all rural people of Bangladesh. Moreover, reluctance, resistance, and time pressure may have influenced some of the respondents when responding to the questionnaire. The result from this study suggest that there are some areas for further research. Further research is essential that would investigate the challenges rural users' experienced during their MFS use. The current study can be replicated with more sample from both rural and urban areas to get in depth understanding about users experience about MFS. Finally, further research is needed to examine strategies that can minimize MFS users' negative experience regarding mobile financial services.

1.7 References

Ahmed, M. T., Imtiaz, M. T., & Kauser, A. A. (2020). A Comparative study of Mobile Banking in Specific Parts of Bangladesh. *International Journal of Science and Business, 4*(6), 129-139. https://doi.org/10.5281/zenodo.3902124

Ahmed, S. M. S., Islam, M. A., Rayhan, S. J., & Mahjabin, S. (2012). Problems and Prospects of Mobile Banking in Bangladesh. *Journal of Arts, Science & Commerce, 3*(1), 47-58.

Akhter, N., & Khalily, M. A. B. (2020). An Analysis of Mobile Financial Services and Financial Inclusion in Bangladesh. *Indian Journal of Human Development, 14*(2), 213–233.https://doi.org/10.1177/0973703020946706

Akhtaruzzaman, M., Islam, M. E., Islam, M. S., Islam, S. R., Bhuiyan, M. R. U., Tareq, M., & Palas, M. J. U. (2017). *An Impact Study on Mobile Financial Services (MFSs) in Bangladesh.* Department of Communications and Publications, Bangladesh Bank. Retrieved September 27, 2020 from https://www.bb.org.bd/pub/special/impact_mfs_27092018.pdf

Alam, M. K. (2017). Understanding Users' Intention to Use Mobile Financial Services: A Study in Khulna City. *Business Review- A Journal of Business Administration Discipline, 12*(1&2), 1-8. http://doi.org/10.35649/KUBR.2017.12.12.1

Alam, M. Z., Patwary, M. M., & Rahim, M. A. (2013). Mobile Money System: The Bangladesh Experience. *International Journal of Scientific and Research Publications, 3*(10), 1-5.

Alkhaldi, A. N. (2017). An empirical examination of customers' mobile phone experience and awareness of mobile banking services in mobile banking in Saudi Arabia. *Interdisciplinary Journal of Information, Knowledge, and Management, 12,* 283-308. https://doi.org/10.28945/3887

Amin, H., & Ramayah, T. (2010). SMS Banking: Explaining the Effects of Attitude, Social Norms and Perceived Security and Privacy. *The Electronic Journal of Information Systems in Developing Countries, 41*(1), 1-15. https://doi.org/10.1002/j.1681-4835.2010.tb00291.x

Bangladesh Bank. (2018). *Bangladesh Mobile Financial Services (MFS) Regulations, 2018*. Payment Systems Department, Bangladesh Bank, Dhaka. Retrieved October 20, 2020 from https://www.bb.org.bd/mediaroom/circulars/psd/jul302018psdl04e.pdf

Bhuiyan, M. S. R., & Rahman, M. M. (2013). Implementation of mobile banking in Bangladesh: Opportunities and Challenges. *IOSR Journal of Electronics and Communication Engineering, 7*(5), 53-58.

Chavali, K., & Kumar, A. (2018). Adoption of Mobile Banking and Perceived Risk in GCC. *Banks and Bank Systems, 13*(1), 72-79. https://doi.org/10.21511/bbs.13(1).2018.07

Chitungo, S. K., & Munongo, S. (2013). Extending the technology acceptance model to mobile banking use in rural Zimbabwe. *Journal of Business Administration and Education*, *3*(1), 51-79.

Chong, A. Y. L. (2013). A two-staged SEM-neural network approach for understanding and predicting the determinants of m-commerce adoption. *Expert Systems with Applications, 40*(4), 1240–1247. https://doi.org/10.1016/j.eswa.2012.08.067

Creswell, J. W. (2012). *Educational Research: Planning, Conducting, and Evaluating Quantitative and Qualitative Research* (4th ed.). Pearson Education, Boston.

Davis, F. D. (1989). Perceived Usefulness, Perceived Ease of Use, and User Acceptance of Information Technology. *MIS Quarterly, 13*(3), 319-340. https://doi.org/10.2307/249008

Dhillon, R. (2014). Mobile Banking in Rural India: Roadmap to Financial Inclusion. *PARIPEX - Indian Journal of Research, 3*(1), 29-31.

Dona, P. D., Mouri, S. M., Hasan, M., & Abedin, M. Z. (2014). Significance of Exponential uses of Mobile Financial Services (MFS) in Bangladesh. *Global Journal of Management and Business Research, 14*(4), 93-101.

Donner, J., & Tellez, C. A. (2008). Mobile banking and economic development: linking adoption, impact, and use. *Asian Journal of Communication, 18*(4), 318–332. https://doi.org/10.1080/01292980802344190

Dube, A., &Helkkula, A. (2015). Service experiences beyond the direct use: indirect customer use experiences of smartphone apps. *Journal of Service Management, 26*(2), 224–248. https://doi.org/10.1108/josm-11-2014-0308

Garg, R., Rahman, Z., & Qureshi, M. N. (2014). Measuring customer experience in banks: scale development and validation. *Journal of Modelling in Management, 9*(1), 87–117. https://doi.org/10.1108/jm2-07-2012-0023

Hasan, M. T. (2020). M-Banking: The Transaction Revolution in Bangladesh. *Science Research, 8*(4), 98-107. https://doi.org/10.11648/j.sr.20200804.12

Helkkula, A., & Kelleher, C. (2010). Circularity of customer service experience and customer perceived value. *Journal of Customer Behaviour, 9*(1), 37–53. https://doi.org/10.1362/147539210x497611

Heinonen, K., Strandvik, T., &Voima, P. (2013). Customer dominant value formation in service. *European Business Review, 25*(2), 104-123. https://doi.org/10.1108/09555341311302639

Hossain, M. S., &Russel, A. H. (2017). Picture of the Socio-Economic Development of Rural People of Bangladesh through Mobile Financial Services - a study on bkash. *European Journal of Business and Management, 9*(36), 78-93.

Islam, M. S., Karia, N., Soliman, M., Fouji, M. H., Khalid, J., & Khaleel, M. (2017). Adoption of Mobile Banking in Bangladesh: A Conceptual Framework. *Review of Social Sciences, 2*(8), 1-8. http://dx.doi.org/10.18533/rss.v2i8.109

Kabir, M. R. (2013). Factors influencing the usage of mobile banking: incident from a developing country. *World Review of Business Research 3*(3), 96-114.

Kabiry, N., & Forghani, M. H. (2013). Factors affecting Isfahanian mobile banking use based on the technology acceptance model. *International Journal of Academic Research in Business and Social Sciences, 3*(5), 611-623.

Karatepe, O. M., & Aga, M. (2013). The effect of job resourcefulness on role stress, emotional exhaustion and overall performance: A study of frontline bank employees. *Journal of Financial Services Marketing, 18*(2), 91–105. https://doi.org/10.1057/fsm.2013.6

Khan, S. N., Akter, M., & Akter, R. (2017). Factors Influencing Adoption and Usage of Mobile Banking: Bangladesh Experience.

International Journal of Finance and Banking Research, 3(1), 1-12. https://doi.org/10.11648/j.ijfbr.20170301.11

Klaus, P. P., & Maklan, S. (2013). Towards a Better Measure of Customer Experience. *International Journal of Market Research, 55*(2), 227–246. https://doi.org/10.2501/ijmr-2013-021

Komulainen, H., & Saraniemi, S. (2019). Customer centricity in mobile banking: a customer experience perspective. *International Journal of Bank Marketing*, 37(5), 1082-1102. https://doi.org/10.1108/IJBM-11-2017-0245

Koksal, M. H. (2016). The intentions of Lebanese consumers to adopt mobile banking. *International Journal of Bank Marketing, 34*(3), 327–346. https://doi.org/10.1108/ijbm-03-2015-0025

Laukkanen, T. (2016). Consumer adoption versus rejection decisions in seemingly similar service innovations: The case of the Internet and mobile banking. *Journal of Business Research, 69*(7), 2432-2439. https://doi.org/10.1016/j.jbusres.2016.01.013

Liza, F. Y. (2014). Factors Influencing the Adoption of Mobile Banking: Perspective Bangladesh. *Global Disclosure of Economics and Business, 3*(2), 199-220.

Luarn, P., & Lin, H.-H. (2005). Toward an understanding of the behavioral intention to use mobile banking. *Computers in Human Behavior, 21*(6), 873–891. https://doi.org/10.1016/j.chb.2004.03.003

Makanyeza, C. (2017). Determinants of consumers' intention to adopt mobile banking services in Zimbabwe. *InternationalJournal of Bank Marketing, 35*(6), 997-1017. https://doi.org/10.1108/IJBM-07-2016-0099

Malaquias, R. F., & Hwang, Y. (2016). An empirical study on trust in mobile banking: A developing country perspective. *Computers in Human Behavior, 54,* 453–461. https://doi.org/10.1016/j.chb.2015.08.039

Malaquias , R. F., & Silva, A. F. (2020). Understanding the use of mobile banking in rural areas of Brazil.*Technology in Society, 62.* https://doi.org/10.1016/j.techsoc.2020.101260

Mbama, C. (2018). Digital banking, customer experience and bank financial performance: UK customers' perceptions. *International Journal of Bank Marketing, 36* (2), 230-255.

Medberg, G., & Heinonen, K. (2014). Invisible value formation: a netnography in retail banking. *International Journal of Bank Marketing, 32*(6), 590-607. http://dx.doi.org/10.1108/IJBM-03-2014-0041

Mohiuddin, G. (2015). A Study on Educational Assessment of Secondary and Higher Secondary Teachers of Bangladesh. *World Scientific News, 16*, 95-115.

Rahman, M., Tazim, M. Z., Das, S., & Islam, L. (2020). State of the art of Mobile Banking Services and Future Prospects in Developing Countries. *9th IEEE International Conference on Communication Systems and Network Technologies*, 145-149. https://doi.org/10.1109/CSNT.2020.28

Ramdhony, D. and Munien, S. (2013). An Investigation on Mobile Banking Adoption and Usage: A Case Study of Mauritius. *World Journal of Social Sciences, 3*(3), 197-217.

Sarpong, N. A., & Agbeko, M. (2020). Optimizing Technology in Mobile Financial Services: User experiences and expectations in Ghana. *International Journal of Computer Applications*, *177*(39), 1-7. http://doi.org/10.5120/ijca2020919716

Shaikh, A., & Karjaluoto, H. (2015). Mobile banking adoption: A literature review. *Telematics and Informatics, 32*(1), 129-142. https://doi.org/10.1016/j.tele.05.003

Sharma, S. K., & Sharma, M. (2019). Examining the role of trust and quality dimensions in the actual usage of mobile banking services: An empirical investigation. *International Journal of Information Management, 44,* 65–75. https://doi.org/10.1016/j.ijinfomgt.2018.09.013

Sultana, B., & Khan, M. M. R. (2016). *Are Mobile Financial Services Promoting Financial Inclusion in Bangladesh? An Assessment Study.* Working Paper Series: WP No 1623, Research Department, Bangladesh Bank. Retrieved October 05, 2020 from http://www.esocialsciences.org/Download/repecDownload.aspx?fname=A2017726145911_20.pdf&fcategory=Articles&AId=11914&fref=repec

Talukder, M. (2014). *Managing Innovation Adoption. From Innovation to Implementation.* Gower Publishing, London, England.https://doi.org/10.4324/9781315593609

Talukder, M., Quazi, A., & Sathye, M. (2014). Mobile Phone Banking Usage Behaviour: An AustralianPerspective. *Australasian Accounting, Business and Finance Journal, 8*(4), 83-104.http://dx.doi.org/10.14453/aabfj.v8i4.6

Tavakol, T. & Dennick, R. (2011). Making Sense of Cronbach's Alpha. *International Journal of Medical Education.* pp. 53-55.

Wei, T. T., Marthandan, G., Chong, A. Y. L., Ooi, K. B., &Arumugam, S. (2009). What drives Malaysian m-commerce adoption? An empirical analysis. *Industrial Management & Data Systems, 109*(3), 370–388.

Wessels, L., &Drennan, J. (2010). An investigation of consumer acceptance of M-banking. *International Journal of Bank Marketing, 28*(7), 547–568. https://doi.org/10.1108/02652321011085194

Zarmpou, T., Saprikis, V., Markos, A., & Vlachopoulou, M. (2012). Modeling users' acceptance of mobile services. *Electronic Commerce Research, 12*(2), 225–248. https://doi.org/10.1007/s10660-012-9092-x

Zhou, T. (2013). An empirical examination of continuance intention of mobile payment services. *Decision Support Systems, 54*(2), 1085–1091. https://doi.org/10.1016/j.dss.2012.10.034

Chapter Two: The Rural Users Challenges with Mobile Financial Services (MFS) in Bangladesh

2.1 Abstract

Mobile Financial Services (MFS) is one of the new inventions of the modern financial system. The study investigated the challenges rural peoples face during their use of MFS in Bangladesh. Specifically, the study sought to gain an understanding of what are their major barriers in using MFS. The study was mainly descriptive research. Data sources were MFS users living in rural areas. The study used a semi-structured interview protocol for rural users. A total of 10 rural users were chosen conveniently from five upazilla in Barishal city. The findings suggest that rural users' low knowledge and awareness level, high transaction cost, and technological issues were some major challenges of users to use MFS. These challenges have an extremely negative impact on their actual use. Therefore, Government and MFS providers need to consider these challenges and take necessary steps to provide a better service to the rural people.

2.2 Introduction

The world is moving forward to the fourth industrial revolution with the continuous development of modern technology. Bangladesh government has also set goals to make "Digital Bangladesh" within 2021. To achieve 'vision 2021' Bangladesh government has introduced mobile financial services in 2011. Mobile financial services (MFS) have made a great opportunity for users to get financial services at any place at any time (Lubna, Bhowmik & Ashraf, 2018).

The rapid growth of mobile phones was visible across different classes of people regardless of their educational qualifications, financial capabilities, and technological backgrounds. According to the Bangladesh Telecommunication Regulatory Commission (BTRC) report, the total number of mobile phone subscribers has reached 161.30 million at the end of June 2020, which is almost 97.94% of the total population. The extraordinary development of the mobile sector all over the world has made a special chance for delivering financial as well as social services through the mobile network (Kabir, 2013). Mobile phones have become a popular tool for everybody, which has created a great opening for the development of banking services through mobile phones to reach the formerly unbanked population (Liza, 2014).

At present, 63% of the people live in the rural areas in Bangladesh (World Bank, 2018), of which most of the people didn't have a bank account (Kabir, Islam & Inam, 2013). In rural areas, formal financial institutions and their services are almost insufficient (Hasan, 2020), because establishing bank branches across rural areas is not an appropriate option due to the costs and regulatory restrictions of the central bank which only grants new branch opening licenses for a maximum of 15 branches in a year (Kabir et al., 2013). Mobile financial service has many positive effects on poor people's lives in numerous ways. Starting from simply sending and receiving money, MFS is used to buy mobile airtime and saves precious time for rural people who would spend on traveling (Islam et al., 2019). Mujeri and Azam (2018) suggest that the MFS providers can reach people living outside mobile phone network coverage locations since it allows people to use mobile phones as an effective distribution channel to lower operational costs, increase financial-services coverage, and extend financial services to un-served populations. Furthermore, as mobile financial service does not require a physical banking facility, so it becomes more useful and inexpensive for rural people (Gupta, Yun, Xu & Kim, 2017). Nevertheless, in emerging economies, the use of mobile devices to execute banking transactions is still low (Upadhyay & Jahanyan, 2016).

Despite developing countries offer huge potential markets for m-commerce service providers, and customers subscribe to mobiles in increasing numbers, the use of mobile banking has not gained grip (Kalinic & Marinkovic, 2016). According to Kabir (2013), the usage of mobile financial services is still a controversial matter among many people because of the cost, availability, and risk involved in such transactions. Additionally, most people do not have a good understanding and abilities about e-banking and/or m-banking. As a result, users have different levels of intention to use mobile financial services (Alkhaldi, 2017). Islam et al. (2019) found that users are also experiencing some problems, for example, security in the transaction, authentication risks, reliability, and connectivity problem between banks and their customers, application security, etc. At present, nearly 50% of the entire rural population of Bangladesh does not use the phone. On the other hand, the user rate is 80% among 16-45 age groups. In rural areas of Bangladesh, only 3% pay bills via mobile, 3% have an online shopping experience and less than 1% earn through online activities (Shadat, Islam, Zahan & Matin, 2020).

Although mobile financial service is safer than digital banking as very few Trojans and viruses exist for phones, mobile financial service is vulnerable to many security threats. Mobile banking users are particularly vulnerable when they receive a fake text message asking for the details of the bank account from a hacker acting as a financial institution. Mobility banking poses security threats such as financial loss, personal data exploitation, cell phone theft, or device loss (Koenig-Lewis, Palmer & Moll, 2010). Gupta and Arora (2017) found that security concerns have noticeably hindered the growth of mobile banking by negatively influencing customers' attitudes to accept it. It is the prime problem in mobile banking services. The intangibility of mobile banking services is the core reason that hinders customers in evaluating services before actual use and enhances their concern over security issues (Makanyeza, 2017). Customers are anxious about the possibility that their money or personal information might be transferred to other parties without consent when transacting via mobile financial services (Ramdhony & Munien, 2013). Security and trust are the most noteworthy aspects of any mobile banking service, but still, people have more doubts about its trustworthiness and security (Dhillon, 2014). Koenig-Lewis et al. (2010) asserted that customer adoption remains the main difficulty that hinders the development of mobile banking.

Since mobile banking is a new experience to most customers, low awareness of technology features is a strong obstacle to its acceptance (Mohammadi, 2015), which causes low customer demand and insufficient returns on investment. Dhillon (2014) found that most of the rural peoples have very little idea about mobile banking and how it is applied and used. It is observed that only 3% of existing customers are using the services of mobile banking in urban areas and they do not have any clear idea about the advantages of mobile banking. It is improbable that banks will increase their investment in mobile banking if customers do not see any advantages in it and if adoption is not extensive (Koenig-Lewis et al., 2010). Moreover, the high quality and continuity of mobile service connectivity are not available in rural areas (Dhillon, 2014).

Mullan, Bradley, and Loane (2017) suggest that the use of a mobile financial service is at an early phase of adoption in several developing countries (Hossain & Haque, 2014) such as Bangladesh, and warrants more research attention (Komulainen & Saraniemi, 2019). People abstain from using mobile banking because it involves payments (Luarn and Lin, 2005), for example, short message service (SMS) fee, security issues, and small keyboards and the display screen of the smartphone. So, understanding what prevents customers from using and adopting mobile banking is very important (Koenig-Lewis et al., 2010). Despite there has been a lot of research conducted on the topic of mobile banking abroad, sadly only a few studies were carried out in Bangladesh (Islam & Salma, 2016). Moreover, few pieces of research have been conducted on mobile financial services especially focusing on the rural people of Bangladesh. As rural people are not so habituated with mobile financial services yet, and they are also experiencing different challenges while using MFS. Therefore, it is very necessary to explore rural users' challenges they face with MFS in Bangladesh.

The purpose of the study is to inquire about the rural mobile financial service users' difficulties they face with MFS. To satisfy the purpose of the study, a research question was formulated. The research particularly investigated the following research question:

1. What are the challenges rural people face at the time of using mobile financial services?

2.3 Methodology

A research design is defined as a plan of research specifying what is to be done and how to do it (Creswell, 2012). This research followed a descriptive research mainly qualitative design.

2.3.1 Location and Sampling

The research was conducted in a small town and village area of the Barishal district of Bangladesh. The population of this study has consisted of mobile financial service users who live in that rural area.

A sample of 10 MFS users was selected for conducting the study. The users were identified from five Upazila (sub-district) which were located in the Barishal district. Five Upazila were selected using convenience sampling considering the accessibility of the researcher. The researcher used convenience sampling considering the researcher's easy access as well as the Upazila proximity to the researcher's place of living. Two MFS users were chosen from each Upazila which means in total 10 MFS users were selected using convenience sampling for interviewing the challenges with MFS. The samples were selected considering easy communication, limitation of time, and financial affordability.

2.3.2 Instrumentation

This study used a semi-structured interview. The interview schedule was employed on 10 MFS users to understand their challenges with MFS. The interview protocols included 6 open-ended questions to explore users' difficulties in using MFS. The data were collected in writing during the interview and a smartphone was used to record the data.

2.3.3 Data Analysis Technique

Data obtained through the interview was analyzed using qualitative approaches for answering the research question. In this study, the qualitative data obtained through the interview was analyzed using the thematic analysis technique. Thematic analysis is a method for identifying and analyzing patterns of meaning/themes in qualitative data (Braun & Clarke, 2014). Thematic analysis results were represented descriptively and narratively.

The sources were coded for analyzing data. In this case, two types of coding were used: alphabetic coding for Upazila and alphanumeric coding for participants. Table 1 shows the coding of the data.

Table 10. Coding of the data source

Categories	**Coding Number**
Upazilas (5)	U_A, U_B, U_C, U_D, U_E
Interview participant Users (10)	IU_1, IU_2, IU_3..............IU_8, IU_9, IU_{10}

2.4 Results and Discussion

The study findings of challenges faced by rural people at the time of using MFS are presented in some major themes below:

2.4.1 Rural users' inadequate knowledge and awareness regarding MFS

The most significant challenge of rural users is inadequate knowledge and low-level awareness about MFS. Rural users with low educational levels have insufficient knowledge about mobile financial services. Although mobile phone usage is increasing among rural people, their MFS usage is still at a very basic level. In Bangladesh, users are increasingly growing the digital lifestyle, but still, most users are not aware of MFS (Hossain & Haque, 2014). Many rural users think MFS can only be used for cash in and cash out. They are not aware of advanced functions of MFS for example, utility bill payment, transfer money to/from a bank or international remittance, and getting balance statement. Bhuiyan and Rahman (2013) found that most of the mobile users in Bangladesh have very little or no education at all. Users with low-level education have difficulties in understanding the MFS menu because of their little reading capabilities. Many rural Bangladeshis have access to basic or feature phones and most USSD menus are in English. It makes it difficult for them to understand or use MFS properly. Some rural users can't even check MFS balance or understand the basic functions of using MFS. Additionally, rural users' low literacy has side effects on the confidence level to use mobile financial services. Ahmed, Imtiaz, and Kauser (2020) further added that many people have almost no knowledge about the most advanced features of mobile banking (Apps). Most of the customers do not know how to write a message on mobile and also most of them are unfamiliar with the English language due to illiteracy (Alam, Patwary & Rahim, 2013). Recently, MFS has introduced the Bengali language in the user interface but still, it's not sufficient for illiterate people to understand and use mobile financial services. According to Ramdhony and Munien (2013), people with having high education are more disposed to use MFS because education shaped their perception of innovations and technologies making them more willing to use MFS.

Three respondents mentioned that the MFS payment process is very complex for them and they need others to help to do that. One of the users $U_C IU_5$ said, "I don't know how to use MFS. My neighbor installed it on my phone. Whenever I got money in the account, I went to the nearest agent for cashing out purposes. They do everything on my behalf." As rural users need other assistance to make the payment or understand basic functions, so they become dependent on others with MFS. Similarly, People were affected by their limited awareness of mobile financial services (Bai, 2019; Mohammadi, 2015) and they can't operate them in a useful manner (Rahman, Tazim, Das & Islam, 2020). If mobile financial service is not as simple as making a phone call or listening to music from a mobile phone, it won't be able to make popular among the mass people in Bangladesh (Bhuiyan & Rahman, 2013). If rural people find MFS easy then it can be assumed that other users will also able to use it comfortably.

2.4.2 High transaction charges of MFS

The study demonstrated that maximum rural users were unhappy due to the high transaction cost of MFS. They mentioned that their biggest challenge is the unbearable transaction cost of MFS. Poor and unbanked rural people can't afford this price. Users running small business thinks the high fees of MFS is a huge burden for them. The previous study also found that most of the users are dissatisfied with the high transaction charges of MFS (Hasan, 2020; Sarpong & Agbeko, 2020). In contrast, Alam et al. (2013) found that most of the users believe mobile banking has a lower cost compared to traditional banking. Most of the respondents mentioned that it will be better to handle cash rather than using it through mobile financial service due to high charges. One respondent $U_A IU_2$ uttered, "If I have ten thousand (10,000.00) taka in my MFS account, I will get one hundred seventy-five (175.00) taka less after the cash-out. So, why would I keep money in an MFS account?" Three respondents believe that MFS is more expensive than traditional banking. Rural and unbanked people belong to the mostly poor or lower-middle-income segment of society. Every penny is valuable for them. Many rural users feel discouraged to use MFS due to its 1.75% cash-out charge. The high fees charged by MFS created a huge barrier for many rural users.

2.4.3 Low level of security and trust concerning MFS

The current worldwide financial crisis made the banks the biggest defaulters. People now have become very unwilling to have unconditional faith in banks and they believe that everything their banks introduce is not safe and secure (Bhuiyan & Rahman, 2013). This study explored that, rural users don't believe in the MFS system fully because of the numerous fraudulent activities that occur with MFS. Similarly, the previous study found that numerous people still do not trust the MFS system as they are concerned about their faulty transactions or frauds (Ahmed et al., 2020; Ramdhony & Munien, 2013). On the contrary, other researchers found that most people believe that mobile banking is fairly or highly secured (Alam et al., 2013; Bhuiyan & Rahman, 2013). Ahmed, Islam, Rayhan & Mahjabin (2012) and Liza (2014) argued that people have high levels of trust across all three perspectives: the banks, mobile service providers, and wireless infrastructure. Many rural users have a problem recalling and correctly enter a PIN. So, they use generic PINs such as 1234. Sometimes they share it with others, including agents, which increases the potential risk of theft. The loss of a user's mobile device often means that criminals can gain access to the

MFS PIN and other sensitive information (Bai, 2019). Criminals use that information to blackmail other users. Bhuiyan and Rahman (2013) suggested that there are concerns about the security of the MFS relating to the confidentiality of the PIN. Most of the time rural users couldn't able to maintain the secrecy of PIN and they become a victim of criminal activities.

Four respondents mentioned that the MFS system cannot be trusted since the system can be breached by hackers. All the respondents mentioned that they fear losing their money from the MFS account because they have heard about numerous fraudulence incidents regarding MFS. Maximum users have safety concerns about using mobile financial services. Three users got the call from fake MFS agents who wanted PINs from them. They think their money may be lost from their MFS account anytime by hackers. A respondent U_EIU_{10} said,

> "Two years ago, I got the call from a fake MFS agent. He told me that I won a pressure cooker as a prize. He told me that I would get a code in a message now, I should tell him. I instantly got a code and told him that code. I never told him my PIN. But, still, I lost twenty thousand (20,000.00) taka from my MFS account. I never got back that money."

Islam et al. (2017) explored that security concern is a major threat that hinders the growth of mobile banking in Bangladesh. The security of MFS has in recent times been questioned as more and more fraudulent activities pass through the MFS system (Sarpong & Agbeko, 2020). Rural people feel unsafe keeping a large amount of money in their MFS accounts. Many rural users were victimized by fraudulent activity and lost their life savings from their MFS account. As the security of MFS service is not up to the mark, so sometimes people lose their account because hacker or fraudulent users can easily hack users account (Rahman et al., 2020). But, Liza (2014) found that all the respondents remained neutral about the security/privacy risk of the mobile banking service. Hossain and Russel (2017) added that MFS has reduced the corruptions and frauds done by dealers, media, and brokers in the rural area.

Three respondents mentioned that there is a high risk of losing their money from the MFS account. One user added that, if someone sends money to the wrong person's account by mistake, then the transaction can't be reversed. It creates high uncertainty with the MFS payments system. Chavali and Kumar (2018) also found that mobile banking is risky during fund transfer because careless mistakes may cause financial loss. It creates more negative effects among users as there is no recovery system for any error (Khan, Akter & Akter, 2017). Rural users are still very amateur with MFS, that's why they may do mistakes during the transaction and lose money. The MFS doesn't allow reversing back transactions if it was made to the wrong account by mistake, so the risk becomes higher and trust level comes to the lower level for rural users.

2.4.4 Harassment of female by male MFS agents

The greatest barrier for rural women to use MFS is the lack of female agents in the community. Women users fear male agents as a result of suffering harassment and eve-teasing from them. Two respondents claimed that their wives' phone numbers were leaked by MFS agents after transactions. Then, their wives got disturbed again and again by those agents. Most agent places are located in crowd place which makes unsafe for female users to go there. Women feel uncomfortable going to a crowded marketplace. A respondent U_AIU_1 said, "Whenever I send money through

MFS, I feel insecure with my wife to collect that money from the nearest agent. She also doesn't like to go to them because the place is very crowded and seems unsafe for her." In Bangladesh, rural women are very traditional-minded. Women users' safety is the primary concern for every rural people. Lubna et al. (2018) suggested that some factors such as user-friendly services, number of branch and agents, rules and regulation, culture, norms, knowledge level, behavioral differences, security, speed, and mobility can ensure women's participation in MFS in Bangladesh. According to Akhtaruzzaman et al. (2017), female participation in the MFS industry of Bangladesh is still small. Because the distributor-agent relationship pattern is still very inflexible, females have to visit nearby markets for taking any mobile banking services, lack of knowledge about the services, etc. Most of the time female users take the help of a male relative who might take advantage of this opportunity to fraud female users. Without female user-friendly service mass women, participation in mobile financial service would be near to impossible.

2.4.5 The inefficiency of customer care services

Sometimes rural users face problems with their MFS which cannot be solved by agents. At that time, they need help from customer care services. Five respondents mentioned that contact with customer care centers is very expensive and difficult. Most of the time the connection remains busy or no customer care representative receives the calls. One respondent claimed that customer care behaves rudely with him as he couldn't able to express his problem clearly to them. Users took a long time to connect with a customer care agent due to a lack of agents and a bad network. Another respondent U_DIU_8 told, "I could able to reach with a customer care representative over the telephone after trying almost thirty (30) minutes. I spoke with him for ten (10) minutes which costs me nearly twenty (20) taka. The cost is huge for me." Akhtaruzzaman et al. (2017) explored that the agents are not competent in dealing with customers' queries and complaints. They do not know how to provide entire services to clients. They are even blamed for colluding with miscreants in illegal transactions through MFS in Bangladesh. Some respondents mentioned that there is no clear guideline about the complaining registration process and where to do that or how much time an MFS provider needs to solve those problems. Sometimes, it takes months for rural users to get any solution for their problems of MFS.

2.4.6 Lack of interoperability among MFS providers

Another huge hindrance of MFS is the lack of interoperability. Almost all the respondents mentioned that the MFS features are very limited. Users can't able to send money among different MFS and they are forced to use multiple MFS to get full efficiency of MFS. Rural users become puzzled in using different brands MFS. It makes the whole situation complex for them. Rural users with bank accounts prefer traditional banking rather than MFS for interoperability among banks. One user U_DIU_7 uttered, "I can send money from my DBBL bank account to any other bank account such as AB bank. But I can't do the same in the case of mobile financial service. Then, why should I use it?" As the MFS market is a concentrated market share across MFS providers, so lack of interoperability in Bangladesh is definite. Most of the users use MFS of the same provider and the low willingness of the dominant MFS creates a lack of interoperability (Mujeri & Azam, 2018). Dominant MFS providers like Bkash don't want interoperability because it won't be profitable for them. Many foreign countries have introduced interoperability which boosted the MFS.

2.4.7 Cash dominated rural society

Rural areas are still cashed dominant and the acceptance level of mobile money is still very low. All the respondents use mainly cash in their day-to-day life. Most of the users mentioned that there are very few places where they can use mobile money at all. In rural areas, everyone accepts cash rather than mobile money. So, saving money in an MFS account seems to be not so useful for them. One of the respondents U_EIU_{10} said, "I cannot buy vegetables or fish or daily necessary products with my mobile money. Small shopkeepers in rural areas also don't accept digital money in some cases. So, I keep money in my money bag rather than MFS account." Rural people have very limited options for buying and selling anything with the digital money of MFS. Moreover, Inam and Islam (2013) mentioned that mobile banking doesn't have a variety of deposit schemes or investment/loan schemes as traditional banking yet. MFS doesn't have much acceptance or diversified schemes like formal banks, therefore rural people rarely prefer to use MFS.

2.4.8 Technological error with MFS

Technological and system error is very common in MFS. Rural users can't make a transaction or take a huge time to do a transaction due to this error. Two respondents mentioned that they couldn't able to send money due to a network problem. One of the users found that transaction time takes too long (more than 5 minutes). Another respondent U_AIU_1 told that "Sometimes, I can't log in to my Bkash account app. Then I need to restart my phone again and again. When I called customer care, they told me it's maybe for some technological difficulty of Bkash." Users can't able to access their MFS account for network problems or compatibility issues of their phone. It means they can't use their own money in need as it's stored in the MFS account. Khan et al. (2017) added that sometimes due to the high load of users, the server got down. Recovery of the server takes some time that leads to a great loss to users as they failed to transact money. The central feature of MFS is its real-time money transfer. Technological problem destroys that feature and makes MFS useless for rural users. Rahman et al. (2020) found that in rural areas, users were unable to utilize mobile banking services because the internet connection is very poor or there might be a poor network connection. In rural area user experience problems with an internet connection and they won't be able to access their account efficiently (Bai, 2019).

2.4.9 Liquidity crisis and distance of nearest MFS agent

Most of the rural users use MFS for cash in and cash out purposes. But, local agents have limited balance which often becomes insufficient for rural users. Five respondents told that during festivals or in any crisis period nearest agent doesn't have enough funds for cash out. When the nearest agent doesn't have enough cash, users need to go a long distance to get their cash. It's not only time-consuming but also expensive for rural users. This study sought that during festivals and crisis period's rural people need to travel a long way to make a payment or cash out money. Likewise, Hasan (2020) found that during Eid or any national ceremony, users of the rural areas face problems to withdraw much money due to insufficient balance of the agents in the company with the availability of agents which is poor in rural areas. One of the respondents U_CIU_6 uttered, "During covid-19 crisis my brother sent me money from another city. Due to the scarcity of cash, I could not able to get money from our nearest agent. So, I had to travel 30 minutes to cash out from MFS." There is a huge scarcity of MFS branches in rural areas. It creates inconvenience for rural users because whenever local agents can't provide any solution for them, they can't reach MFS branches.

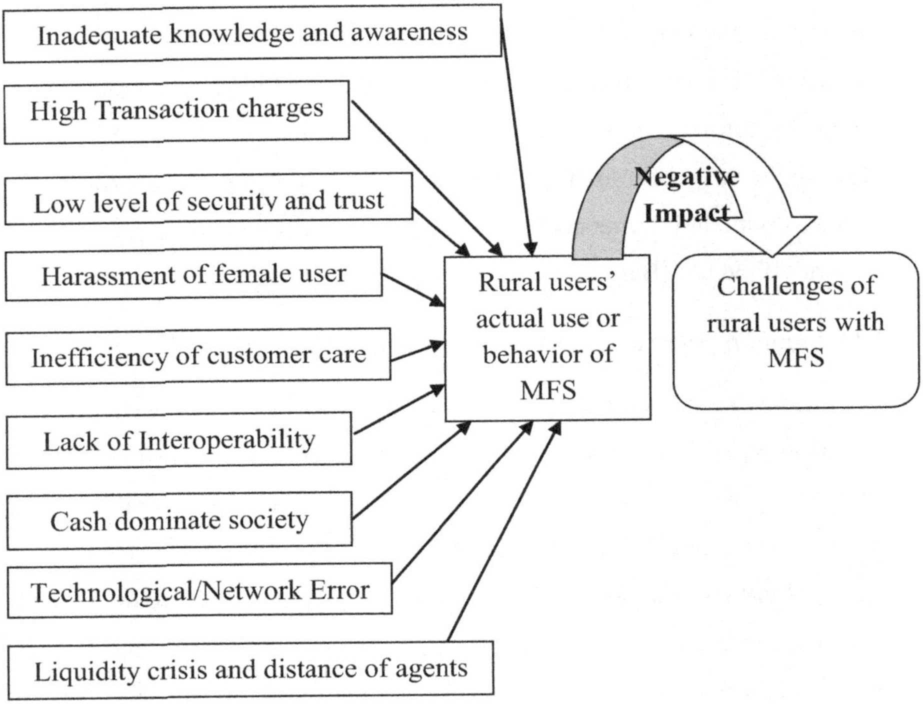

Figure 7. The negative impact of rural users challenges on their actual behavior

This research has established that several issues are causing challenges for rural users during their use of mobile financial services. Rural users' scarce knowledge and awareness, incompetence of customer care agents, harassment of female users; cash dominate society, lack of interoperability, huge transaction fees, low level of security, trust, and technological error of MFS are the core challenges for people to use MFS. MFS providers are mainly blamed due to generating these challenges for rural people. Challenges of rural users with MFS have created a negative impact on their actual use of MFS. Many rural users are still out of the reach for MFS due to these challenges. A good number of rural people have given up using MFS because of the never-ending difficulties with MFS. So, MFS providers and government must come forward to find necessary ways to reduce these challenges of rural users with MFS.

2.5 Conclusion and Managerial Implication

MFS is a fast-growing business. Bangladesh has introduced MFS to reduce poverty especially in rural areas and ensure financial inclusion all over the country. So, MFS providers need to make some effective development to the quality of their service to ensure smooth transactions in the rural area. MFS providers would able to design their service only after understanding rural users' difficulties with it.

This study has explored rural users' core challenges with MFS and its negative impact on their actual use. It can be used as a point of reference for the government in bringing the unbanked rural population under formal banking channels through increasing the use of mobile financial services. This study would help the central bank in making necessary policy and providing helpful guidelines to all stakeholders so that MFS can become more prospective for the rural disadvantageous people.

Mobile financial service is an emerging sector and there exist huge potentials for the whole economy of Bangladesh. MFS makes formal banking more convenient and accessible for users as it offers real-time online banking, available anytime, anywhere throughout the country. It has covered the whole country within a short time targeting low-income and unbanked populations. At present people get modern financial transaction facilities through their MFS which is a very attractive feature than the traditional banking system. There is increasing growth in the acceptance of mobile financial services all over the country. But in reality, MFS is very much new for rural users. The degree of awareness for MFS is not up to the mark yet. The safety-security and affordability of MFS have negatively impacted rural peoples' MFS use. Proper methodologies need to be identified to show rural users how MFS facilitates access to banking services for them. The reduction of risk associated with MFS can certainly reduce challenges as well as may prove to be an important innovation for developing countries like Bangladesh.

2.6 References

Ahmed, M. T., Imtiaz, M. T., & Kauser, A. A. (2020). A Comparative study of Mobile Banking in Specific Parts of Bangladesh. *International Journal of Science and Business, 4*(6), 129-139. https://doi.org/10.5281/zenodo.3902124

Ahmed, S. M. S., Islam, M. A., Rayhan, S. J., & Mahjabin, S. (2012). PROBLEMS AND PROSPECTS OF MOBILE BANKING IN BANGLADESH. *Journal of Arts, Science & Commerce, 3*(1), 47-58.

Akhtaruzzaman, M., Islam, M. E., Islam, M. S., Islam, S. R., Bhuiyan, M. R. U., Tareq, M., & Palas, M. J. U. (2017). *An Impact Study on*

Mobile Financial Services (MFSs) in Bangladesh. Department of Communications and Publications, Bangladesh Bank. Retrieved September 27, 2020 from https://www.bb.org.bd/pub/special/impact_mfs_27092018.pdf

Alam, M. Z., Patwary, M. M., & Rahim, M. A. (2013). Mobile Money System: The Bangladesh Experience. *International Journal of Scientific and Research Publications, 3*(10), 1-5.

Alkhaldi, A. N. (2017). An empirical examination of customers' mobile phone experience and awareness of mobile banking services in mobile banking in Saudi Arabia. *Interdisciplinary Journal of Information, Knowledge, and Management, 12,* 283-308. https://doi.org/10.28945/3887

Bai, M. H. (2019). Mobile Banking Services and Customer Satisfaction with Reference to ICICI Bank - A Study. *Shanlax International Journal of Commerce, 7*(2), 7–18. https://doi.org/10.34293/commerce.v7i2.345

Bhuiyan, M. S. R., & Rahman, M. M. (2013). Implementation of mobile banking in Bangladesh: Opportunities and Challenges. *IOSR Journal of Electronics and Communication Engineering, 7*(5), 53-58.

Braun, V., & Clarke, B. (2014). What can "thematic analysis" offer health and wellbeing researchers? *International Journal of Qualitative Studies on Health and Wellbeing, 9*:1, 26152. http://dx.doi.org/10.3402/qhw.v9.26152

Chavali, K., & Kumar, A. (2018). Adoption of Mobile Banking and Perceived Risk in GCC. *Banks and Bank Systems, 13*(1), 72-79. https://doi.org/10.21511/bbs.13(1).2018.07

Creswell, J. W. (2012). *Educational Research: Planning, Conducting, and Evaluating Quantitative and Qualitative Research* (4th ed.). Pearson Education, Boston.

Dhillon, R. (2014). Mobile Banking in Rural India: Roadmap to Financial Inclusion. *PARIPEX - Indian Journal of Research, 3*(1), 29-31.

Gupta, A., & Arora, N. (2017). Understanding determinants and barriers of mobile shopping adoption using behavioral reasoning theory. *Journal of Retailing and Consumer Services, 36,* 1-7. https://doi.org/10.1016/j.jretconser.2016.12.012

Gupta, S., Yun, H., Xu, H., & Kim, H. (2017). An exploratory study on mobile banking adoption in Indian metropolitan and urban areas: a scenario-based experiment. *Information Technology for Development, 23*(1), 127–152. https://doi.org/10.1080/02681102.2016.1233855

Hasan, M. T. (2020). M-Banking: The Transaction Revolution in Bangladesh. *Science Research, 8*(4), 98-107. https://doi.org/10.11648/j.sr.20200804.12

Hossain, M. A., & Haque, M. Z. (2014). Prospects and Challenges of Mobile Banking in Bangladesh. *Journal of Business Studies, 35*(2), 165-186.

Hossain, M. S., & Russel, A. H. (2017). Picture of the Socio-Economic Development of Rural People of Bangladesh through Mobile

Financial Services - a study on bkash. *European Journal of Business and Management, 9*(36), 78-93.

Islam, N., Mustafi, M. A. A., Rahman, M. N., Nower, N., Rafi, M. M. A., Natasha, M. T., Hassan, R., & Afrin, S. (2019). Factors Affecting Customers' Experience in Mobile Banking of Bangladesh. Global *Journal of Management and Business Research: A Administration and Management, 19*(5), 36-49. https://doi.org/10.34257/GJMBRAVOL19IS5PG37

Inam, M. T., & Islam, M. B. (2013). Possibilities and Challenges of Mobile Banking: A case study in Bangladesh. *International Journal of Advanced Computational Engineering and Networking, 1*(3), 20-24.

Islam, M. S., Karia, N., Soliman, M., Fouji, M. H., Khalid, J., & Khaleel, M. (2017). Adoption of Mobile Banking in Bangladesh: A Conceptual Framework. *Review of Social Sciences, 2*(8), 1-8. http://dx.doi.org/10.18533/rss.v2i8.109

Islam, K. M. A., & Salma, U. (2016). Mobile Banking Operations and Banking Facilities to Rural People in Bangladesh. *International Journal of Finance and Banking Research, 2*(4), 147-162. http://dx.doi.org/10.11648/j.ijfbr.20160204.14

Kabir, M. R. (2013). Factors influencing the usage of mobile banking: incident from a developing country. *World Review of Business Research 3*(3), 96-114.

Kabir, M. E., Islam, M. B., & Inam, M. T. (2013). Uniform Payment System for Banking Industries: A Case Study in Bangladesh. *International Journal of Computer and Electronics Research, 2*(2), 81-88.

Kalinic, Z., & Marinkovic, V. (2016). Determinants of users' intention to adopt m-commerce: An empirical analysis. *Information Systems and e-Business Management, 14*(2), 367-387. https://doi.org/10.1007/s10257-015-0287-2

Khan, S. N., Akter, M., & Akter, R. (2017). Factors Influencing Adoption and Usage of Mobile Banking: Bangladesh Experience. *International Journal of Finance and Banking Research, 3*(1), 1-12. https://doi.org/10.11648/j.ijfbr.20170301.11

Koenig-Lewis, N., Palmer, A., & Moll, A. (2010). Predicting young consumers' take up of mobile banking services. *International Journal of Bank Marketing, 28*(5), 410–432. https://doi.org/10.1108/02652321011064917

Liza, F. Y. (2014). Factors Influencing the Adoption of Mobile Banking: Perspective Bangladesh. *Global Disclosure of Economics and Business, 3*(2), 199-220.

Luarn, P., & Lin, H.-H. (2005). Toward an understanding of the behavioral intention to use mobile banking. *Computers in Human Behavior, 21*(6), 873–891. https://doi.org/10.1016/j.chb.2004.03.003

Lubna, T., Bhowmik, S., & Ashraf, S. (2018). Factors Influencing Women's Access of Mobile Financial Services: An Empirical Study on Sylhet Region. *European Journal of Business and Management, 10*(8), 33-40. https://doi.org/10.7176/EJBM

Makanyeza, C. (2017). Determinants of consumers' intention to adopt mobile banking services in Zimbabwe. *International Journal of Bank Marketing, 35*(6), 997-1017. https://doi.org/10.1108/IJBM-07-2016-0099

Mohammadi, H. (2015). A study of mobile banking loyalty in Iran. *Computers in Human Behavior, 44,* 35–47. http://dx.doi.org/10.1016/j.chb.2014.11.015

Mujeri, M. K., & Azam, S. E. (2018). *Interoperability of Digital Finance in Bangladesh: Challenges and Taking-Off Options.* Working Paper No. 54, Institute for Inclusive Finance and Development. Retrieved October 05, 2020 from http://inm.org.bd/wp-content/uploads/2018/06/Working-Paper-54.pdf

Mullan, J., Bradley, L., & Loane, S. (2017). Bank adoption of mobile banking: stakeholder perspective. *International Journal of Bank Marketing, 35*(7), 1154-1174. https://doi.org/10.1108/IJBM-09-2015-0145

Rahman, M., Tazim, M. Z., Das, S., & Islam, L. (2020). State of the art of Mobile Banking Services and Future Prospects in Developing Countries. *9th IEEE International Conference on Communication Systems and Network Technologies*, 145-149. https://doi.org/10.1109/CSNT.2020.28

Ramdhony, D. and Munien, S. (2013). An Investigation on Mobile Banking Adoption and Usage: A Case Study of Mauritius. *World Journal of Social Sciences, 3*(3), 197-217.

Sarpong, N. A., & Agbeko, M. (2020). Optimizing Technology in Mobile Financial Services: User experiences and expectations in

Ghana. *International Journal of Computer Applications*, *177*(39), 1-7. http://doi.org/10.5120/ijca2020919716

Shadat, W. B., Islam, M. S., Zahan, I., & Matin, M. (2020). Ep 1: Digital Literacy in Rural Bangladesh. Webinar Deck. *Brac Institute of Governance & Development.* Retrieved October 15, 2020 from https://bigd.bracu.ac.bd/wp-content/uploads/2020/09/Webinar-Deck-Digital-Literacy-in-Rural-Bangladesh.pdf

Upadhyay, P., & Jahanyan, S. (2016). Analyzing user perspective on the factors affecting use intention of mobile based transfer payment. *Internet Research. 26*(1), 38–56. https://doi.org/10.1108/IntR-05-2014-0143

World Bank. (2018).*World Development Indicators: Rural environment and land use.* Retrieved October 05, 2020 from http://wdi.worldbank.org/table/3.1#

Publisher: Eliva Press Global Ltd

Email: info@elivapress.com